MW00389763

ABC Tracing Book For Kids

Letter Tracing Practice Workbook for Toddlers and Preschoolers

PRESCHOOL ACTIVITY BOOKS

ISBN-13: 978-1976002946
ISBN-10: 197600294X

THIS PRACTICE WORKBOOK FOR PRESCHOOLERS IS PERFECT FOR HOMESCHOOL, PRESCHOOL, AND KINDERGARTEN. PRACTICING HANDWRITING CAN HELP YOUNG CHILDREN TO DEVELOP THE FINE MOTOR CONTROL THEY NEED FOR THE FUTURE. THIS WORKBOOK WILL HELP YOUR TODDLER OR PRESCHOOL AGE CHILD LEARN THE ALPHABET, IMPROVE THEIR HANDWRITING SKILLS AND FAMILIARIZE THEM WITH 100 SIGHT WORDS AS THEY PRACTICE WRITING!

WHAT'S INSIDE:

- 108 PAGES OF PRACTICE
- 4 PAGES PER LETTER INCLUDING ONE FULL BLANK WRITING PAGE (PER LETTER) FOR YOUR CHILD TO PRACTICE WITHOUT TRACING
- LARGE 8 X 10 INCH PAGES
- FUN PICTURES TO COLOR FOR EVERY LETTER
- 100 SIGHT WORDS TO PRACTICE WRITING

HERE'S A TIP:
TO EXTEND THE USE OF THIS BOOK REMOVE THESE PAGES AND PUT THEM IN PLASTIC SHEET PROTECTORS. PLACE THE COVERED PAGES INTO A BINDER AND HAVE YOUR CHILD USE A DRY ERASE MARKER TO PRACTICE AND WIPE AWAY WHEN DONE. NOW, YOUR CHILD CAN PRACTICE THEIR HANDWRITING ON AN UNLIMITED AMOUNT OF PAGES!

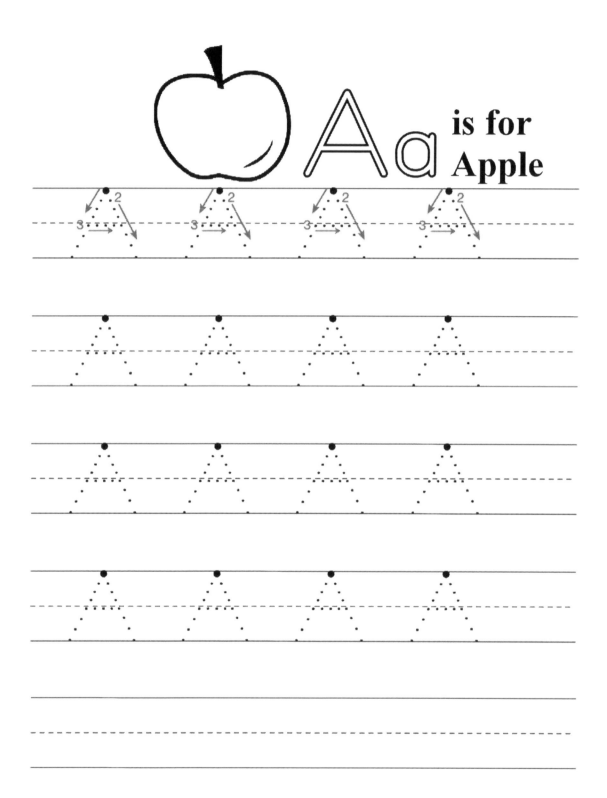

Aa is for
Apple

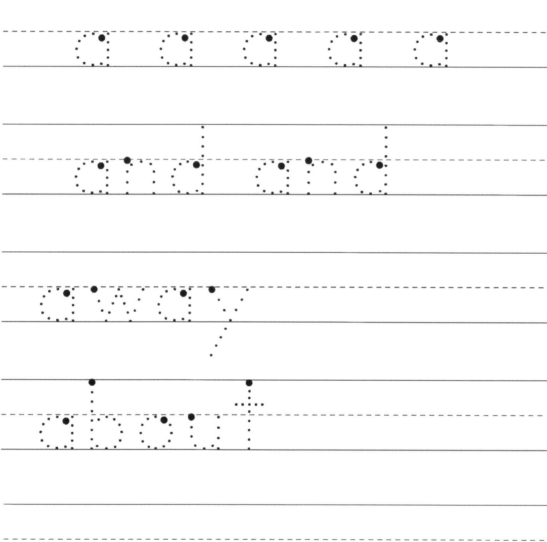

a a a a a

and and

away

about

B b is for
Butterfly

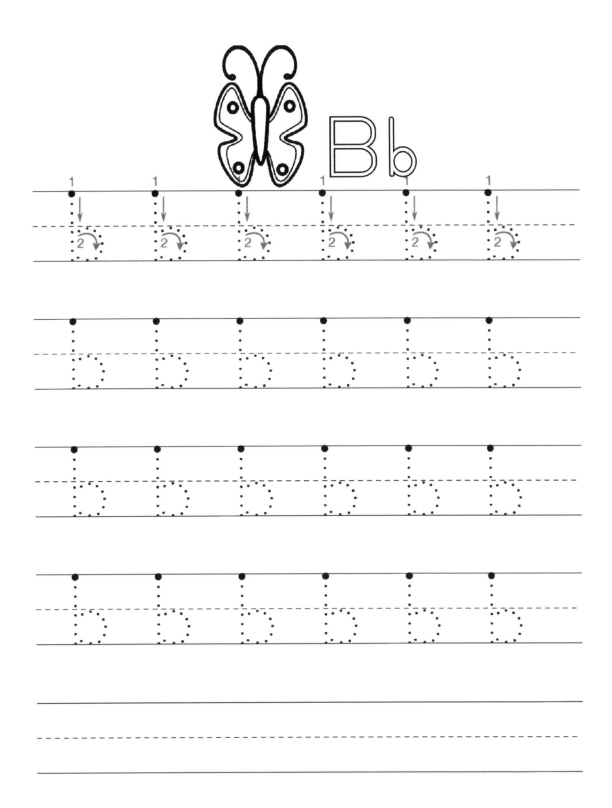

Bb

be be be

big big big

blue

brown

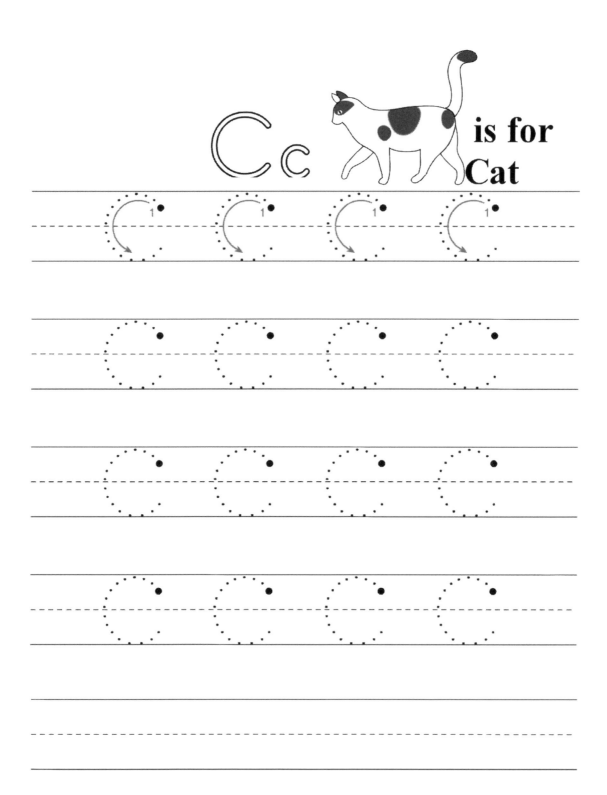

C c is for Cat

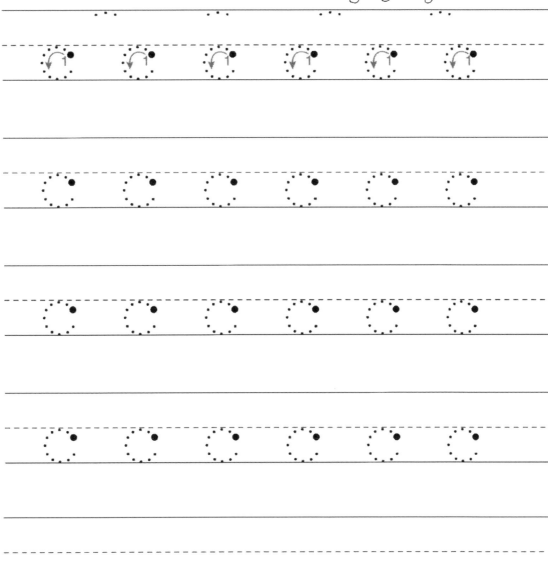

Cc

cat cat

cat cat

come

come

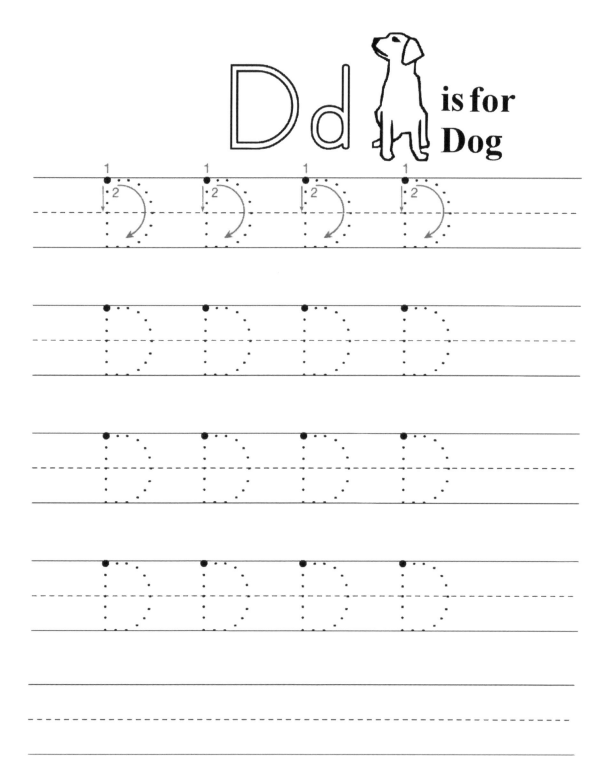

D d is for Dog

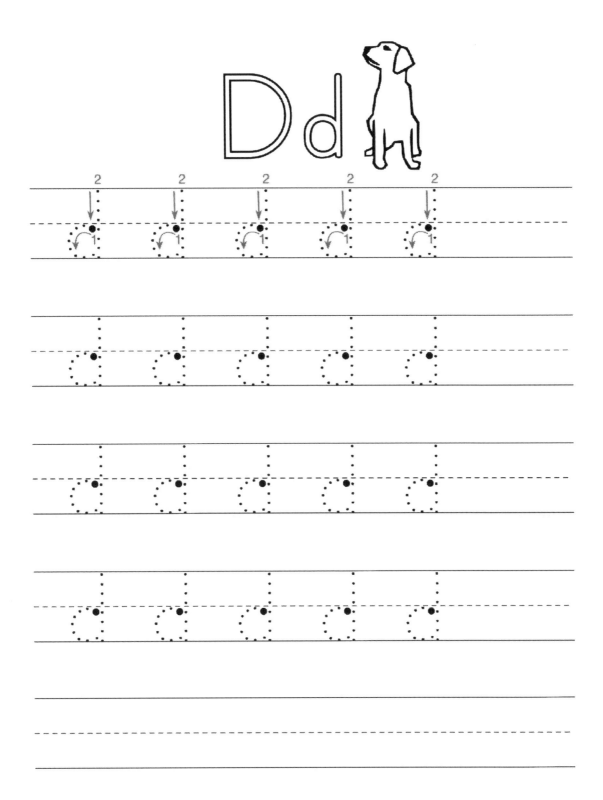

D d

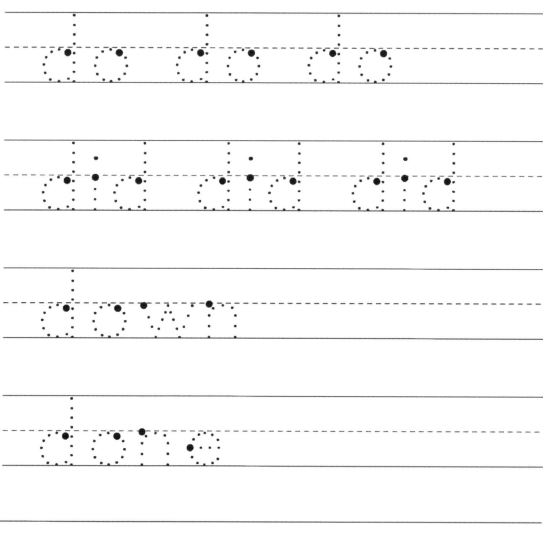

do do do

did did did

down

done

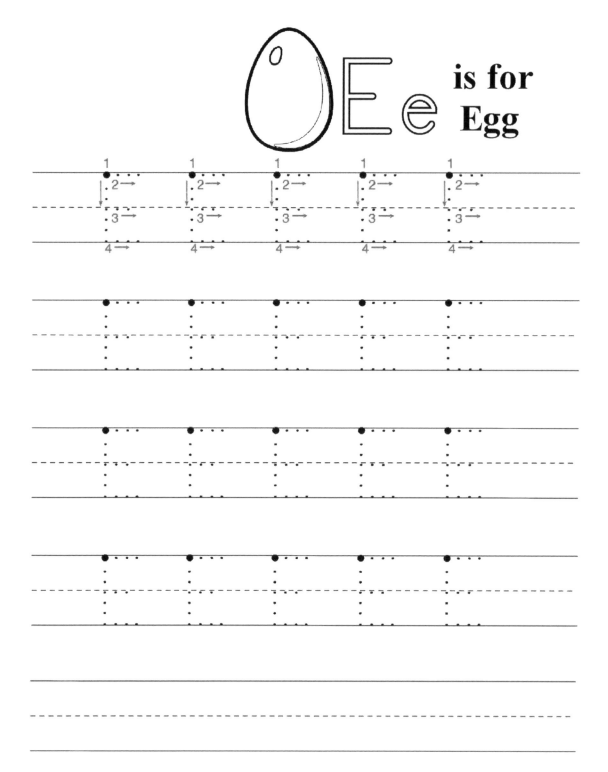

is for
Egg

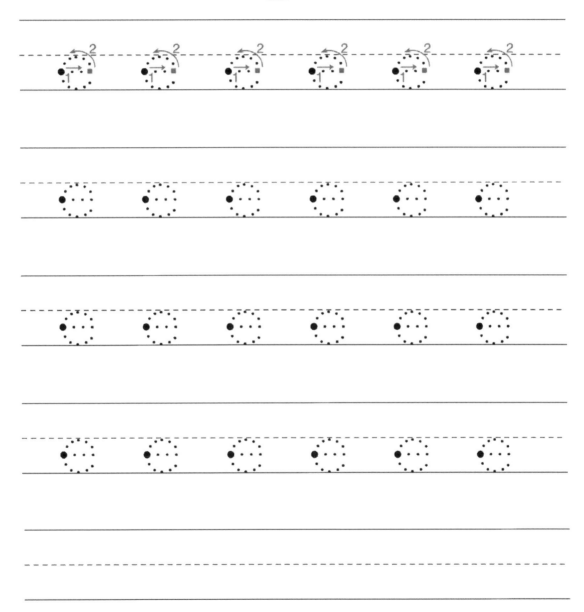

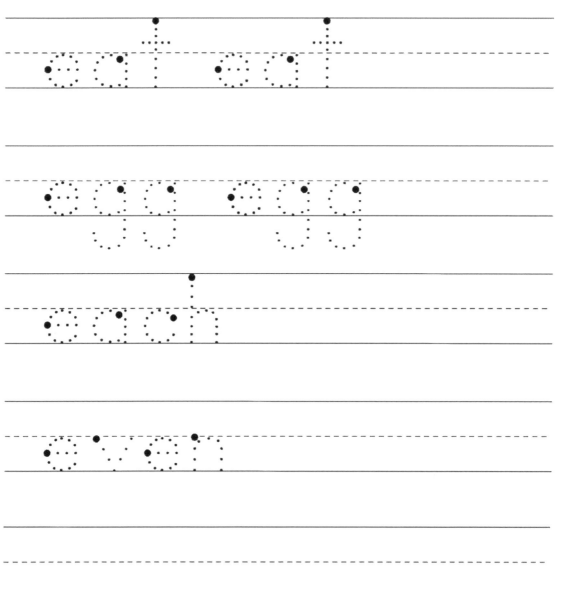

eat eat

egg egg

each

even

F f

is for
Flower

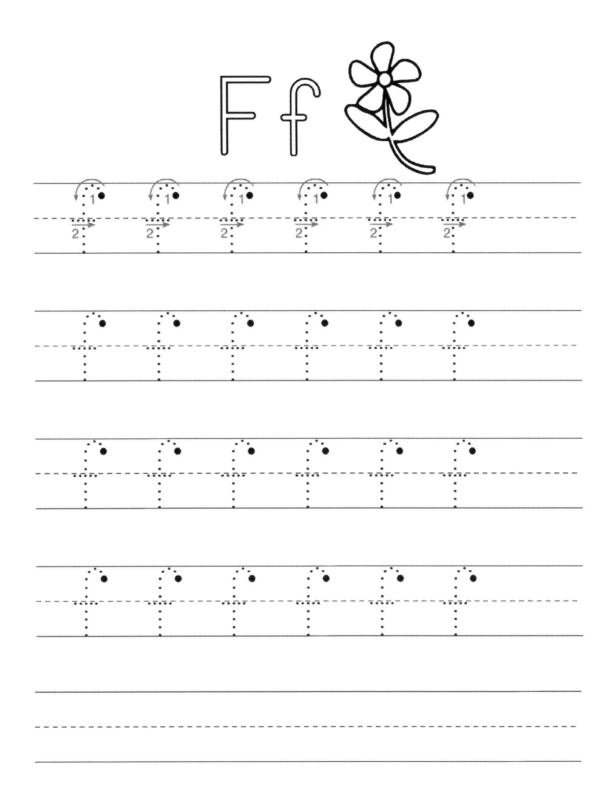

F f

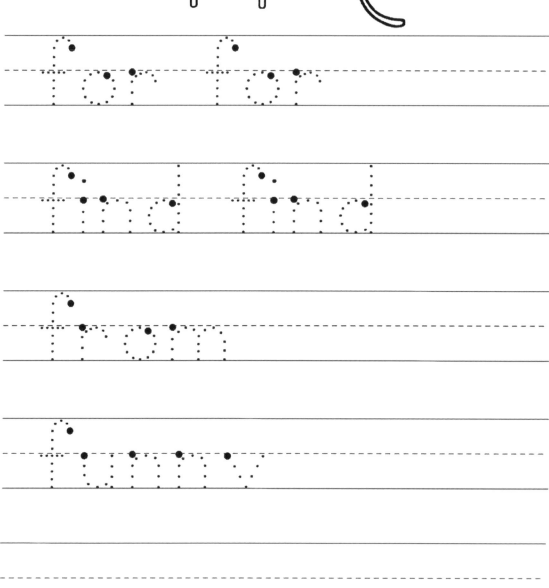

for for for

find find

from

funny

G g

is for
Goat

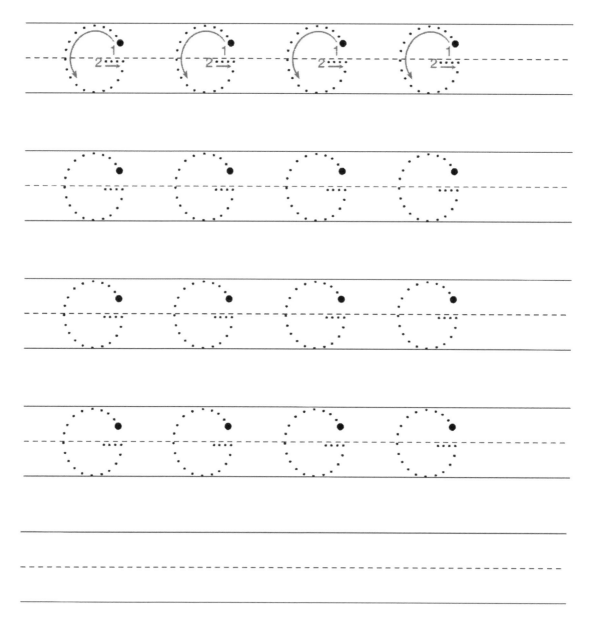

G g

Gg

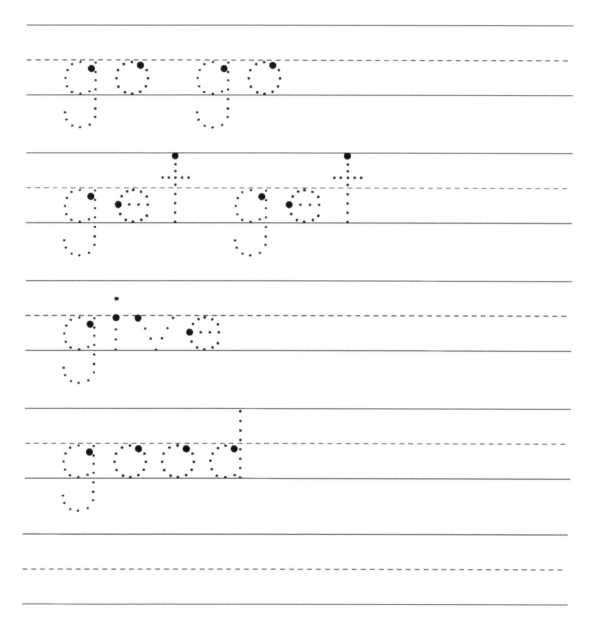

go go

get get

give

good

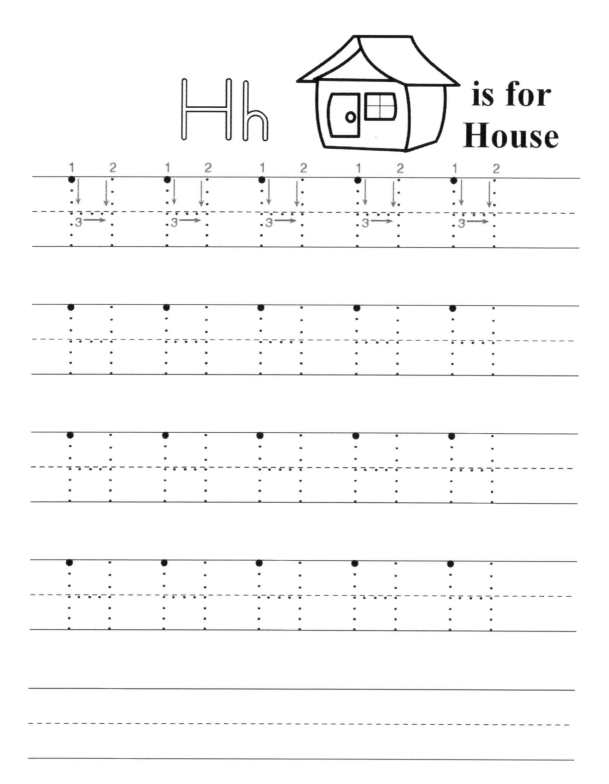

Hh

is for
House

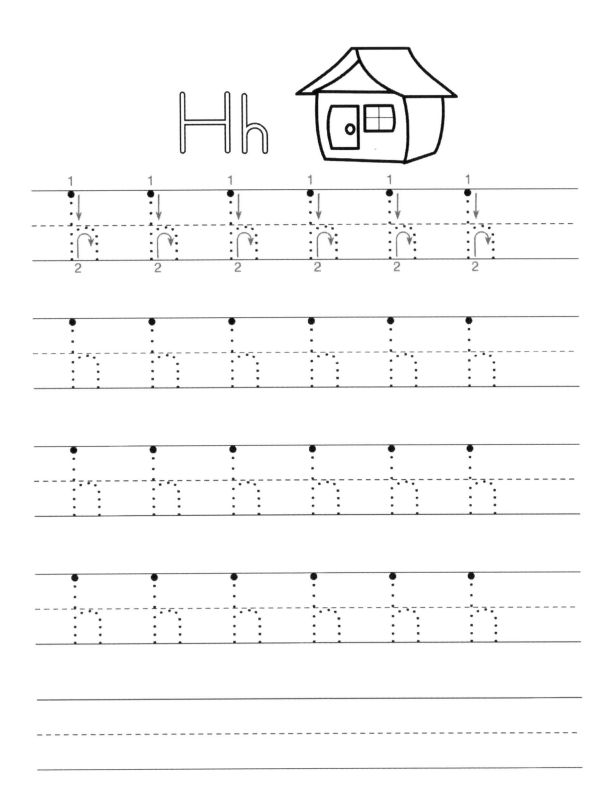

Hh

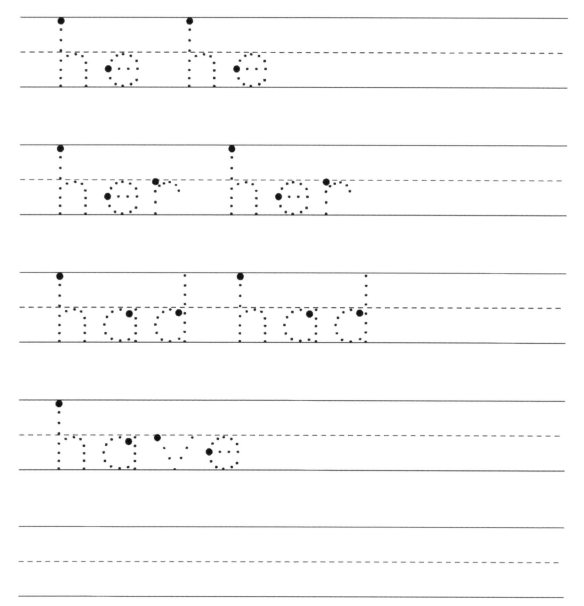

he he

her her

had had

have

Ii is for Island

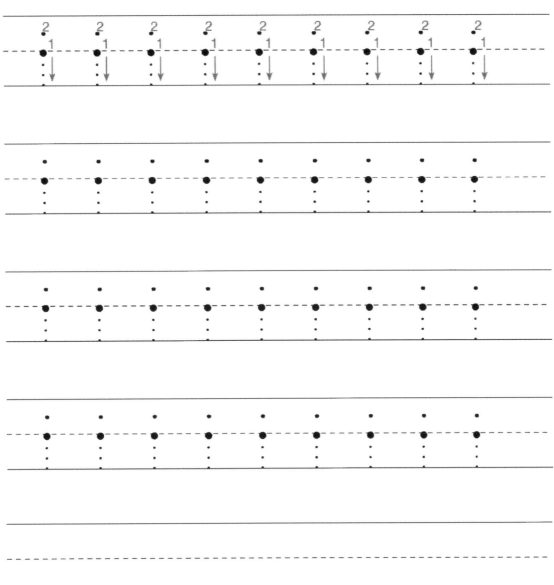

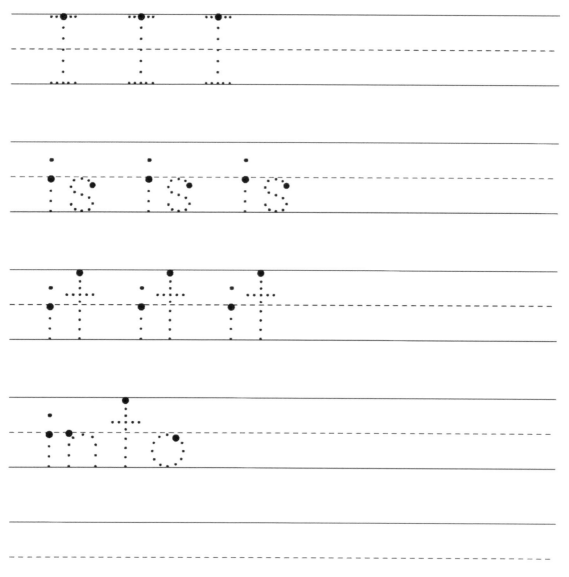

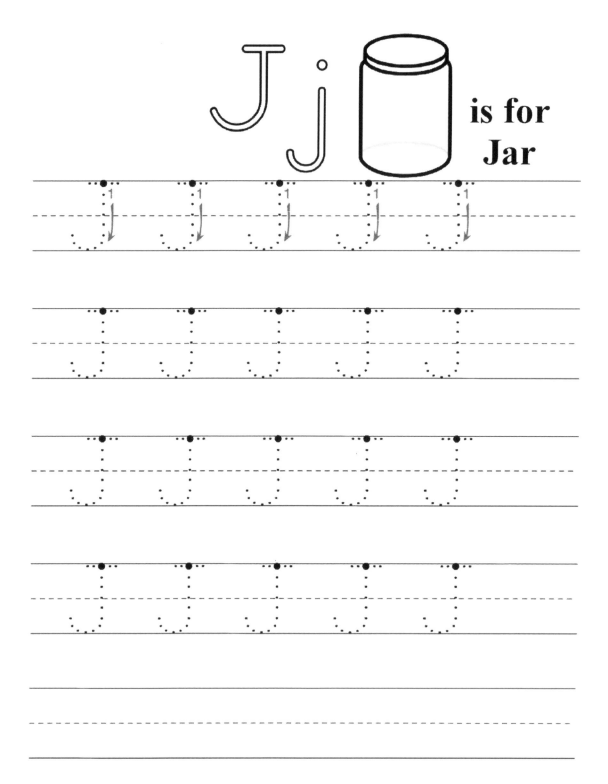

J j is for Jar

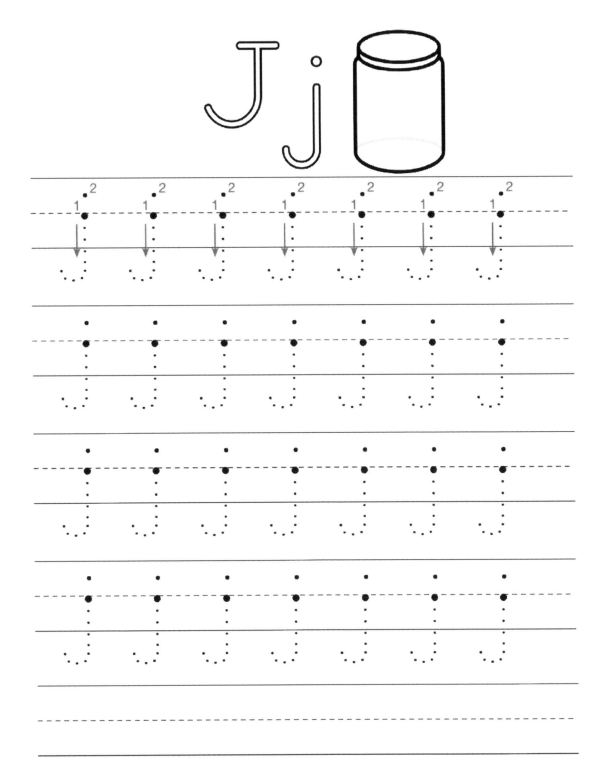

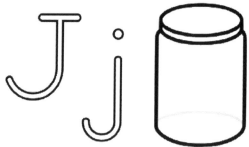

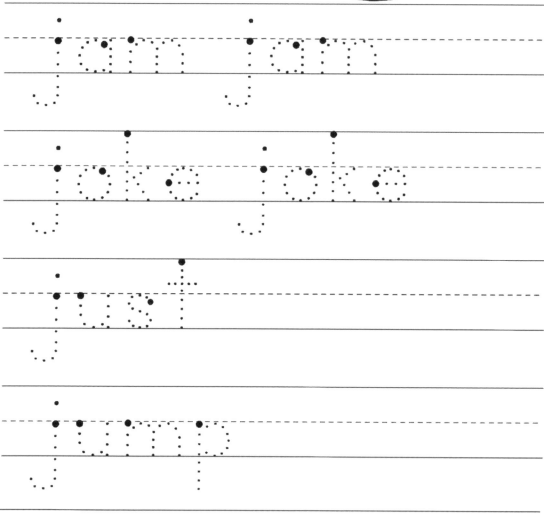

jam jam

joke joke

just

jump

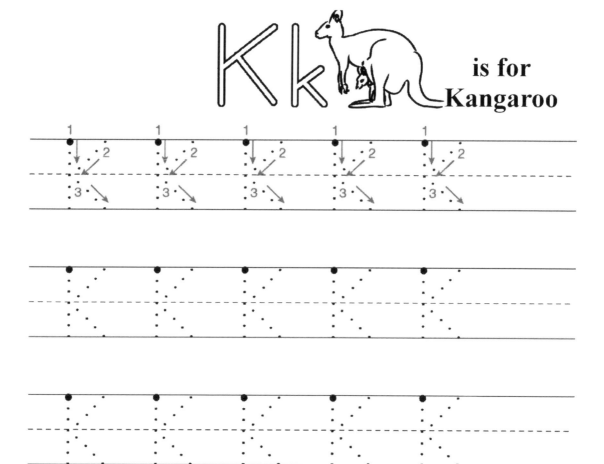

is for
Kangaroo

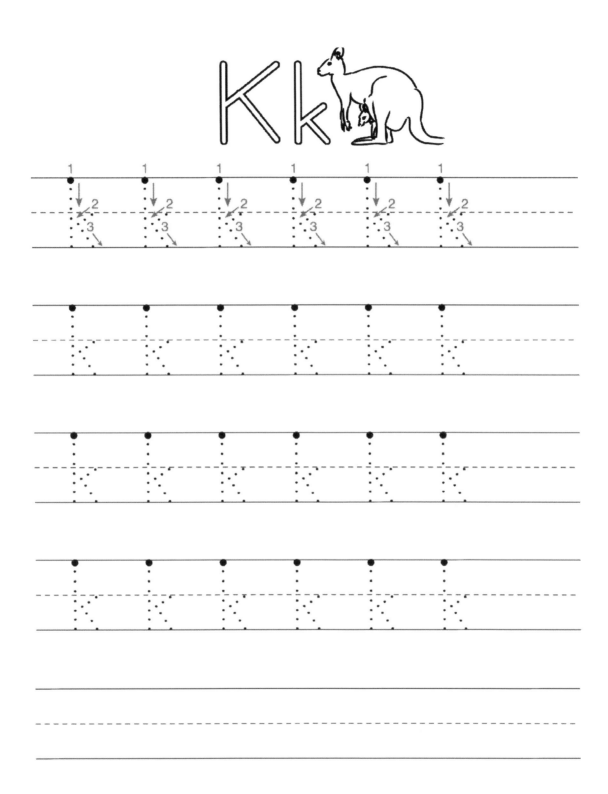

K k

kid kid

kite kite

know

kind

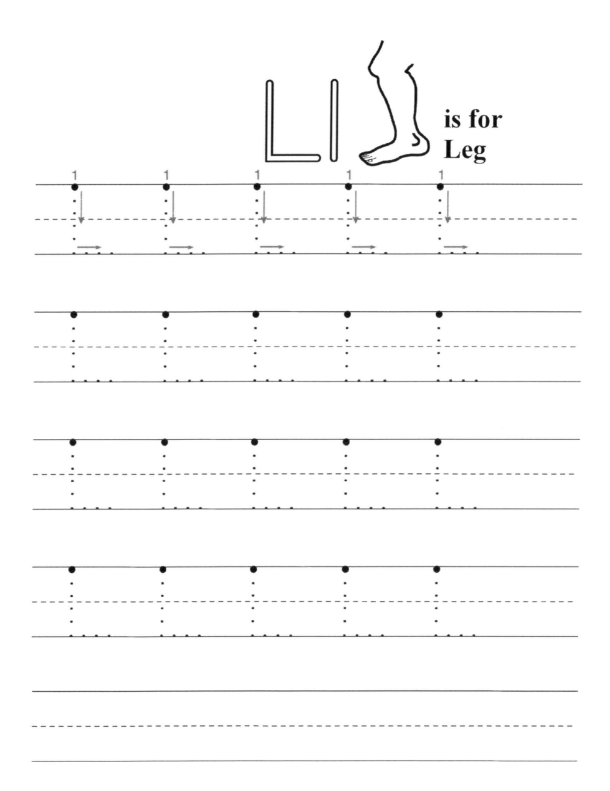

L l is for Leg

Ll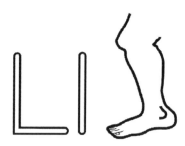

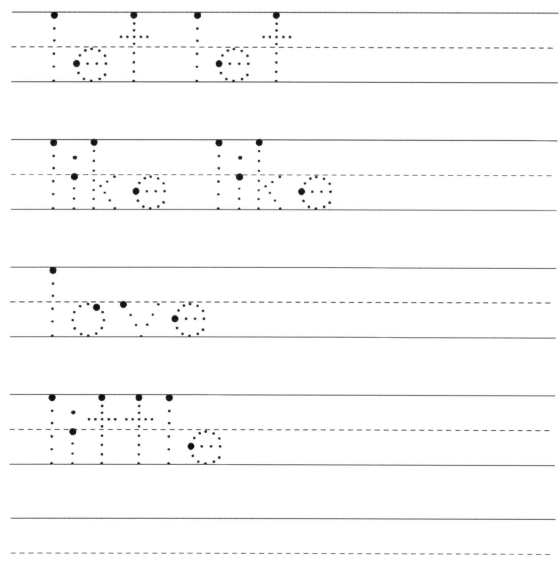

let let

like like

love

little

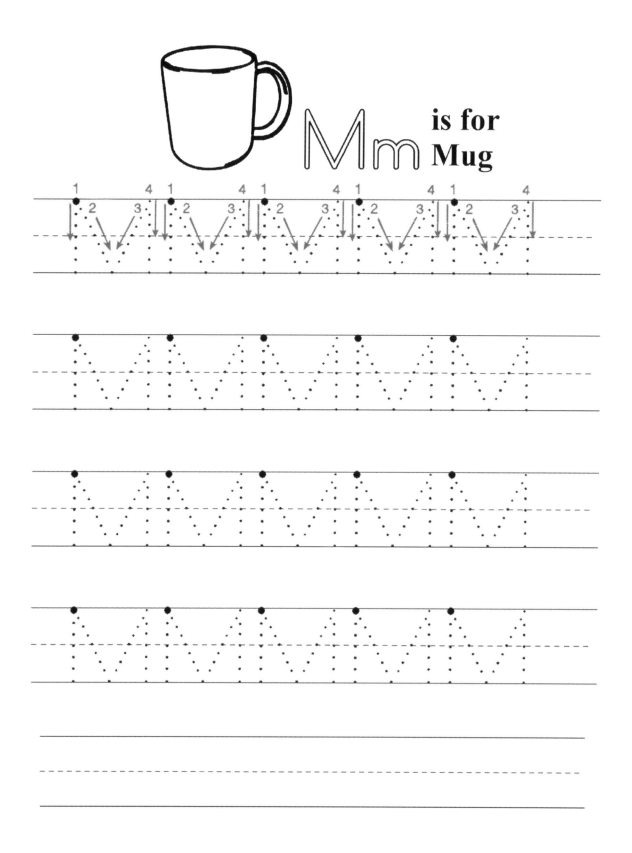

M m is for Mug

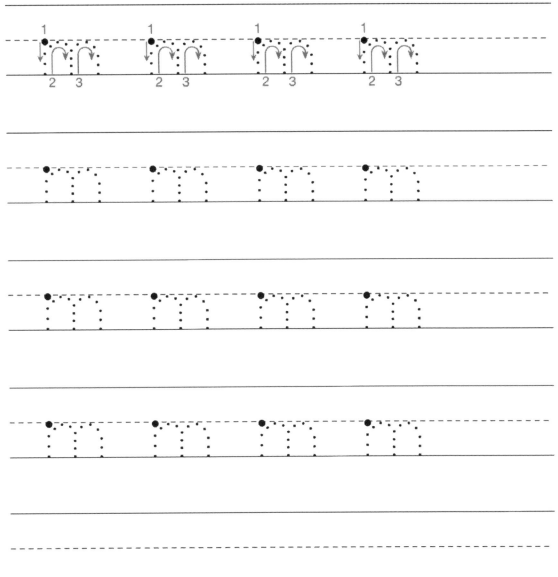

Mm

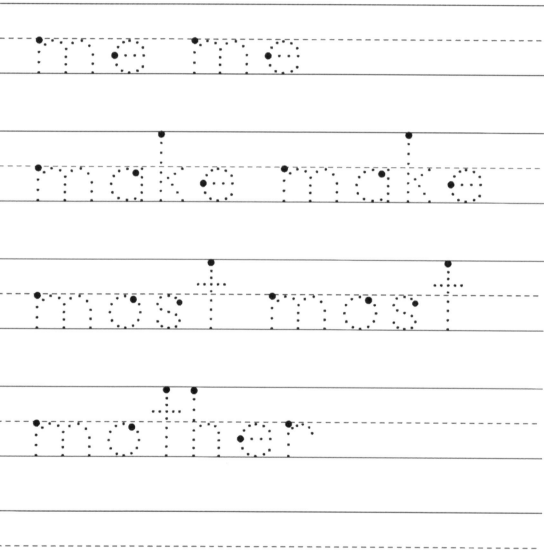

me me

make make

most most

mother

Nn is for Notebook

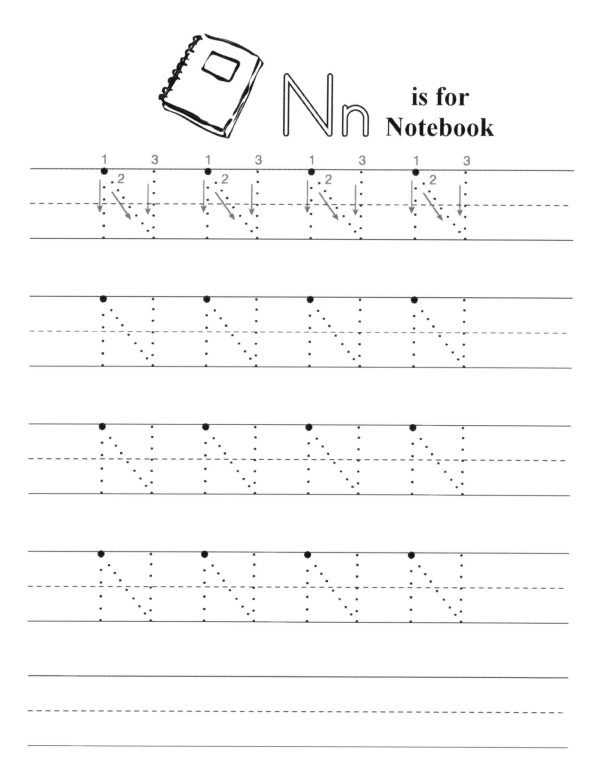

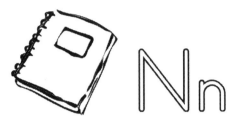

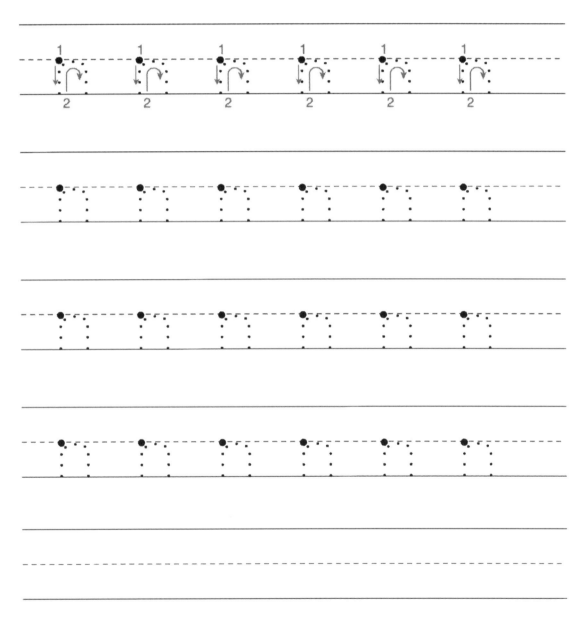

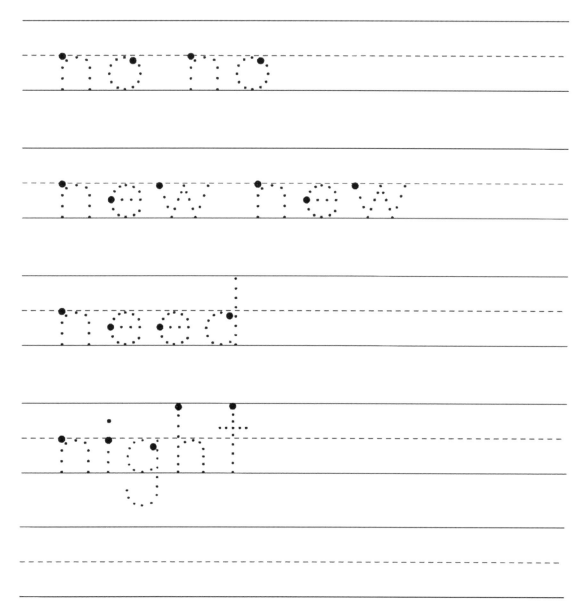

is for
Owl

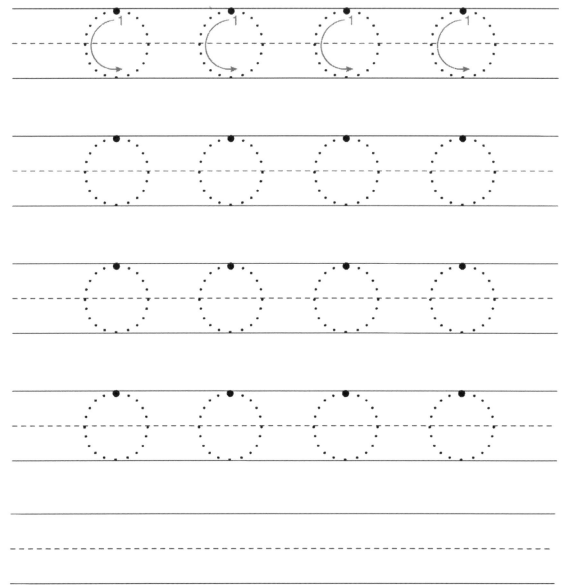

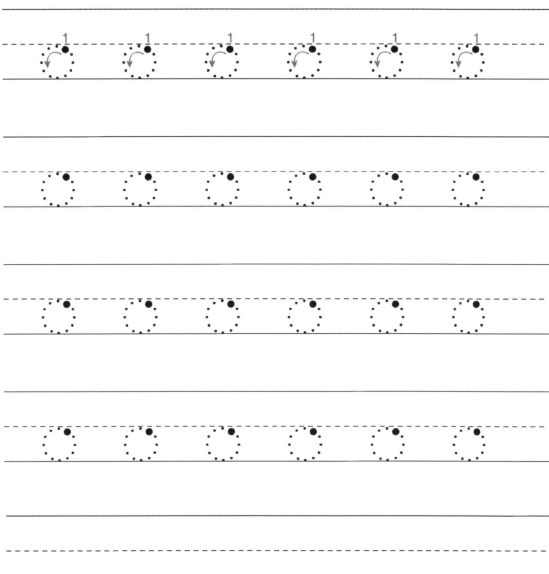

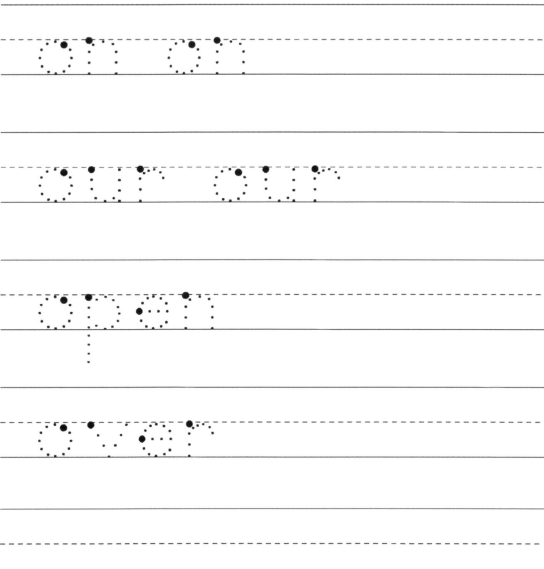

on on

our our

open

over

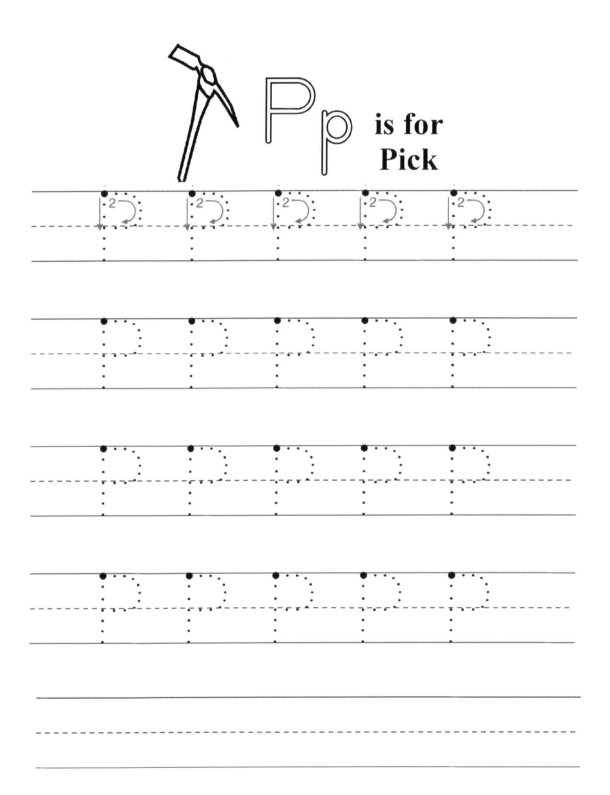

is for
Pick

Pp

1 1 1 1 1 1

P P P P P P

P P P P P P

P P P P P P

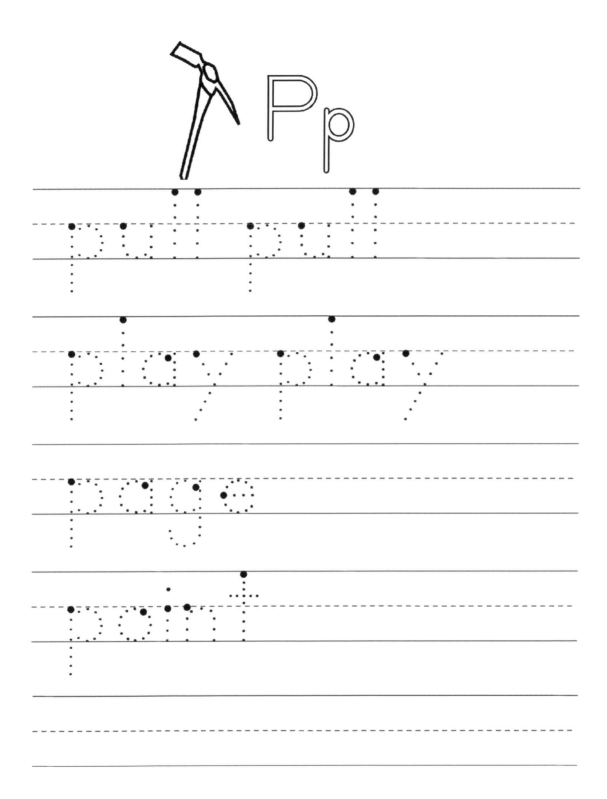

Pp

pull pull

play play

page

point

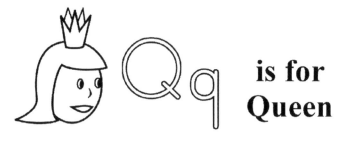

is for Queen

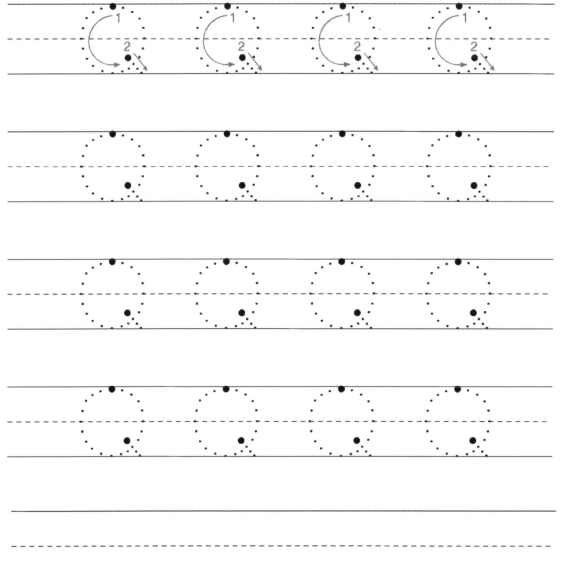

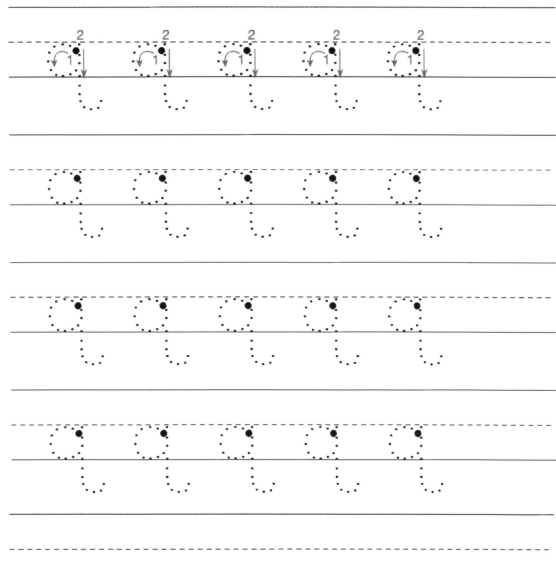

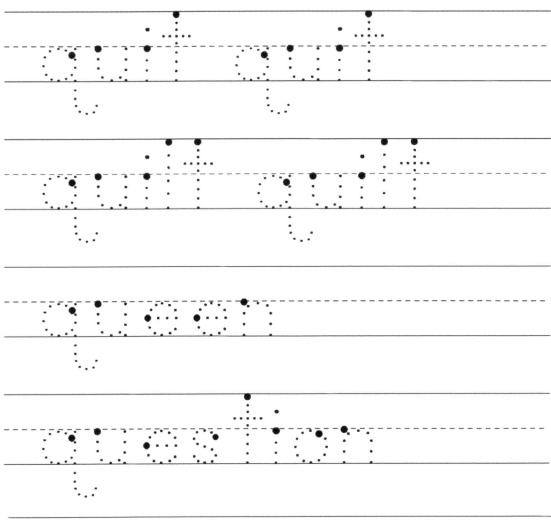

quit quit

quit quit

queen

question

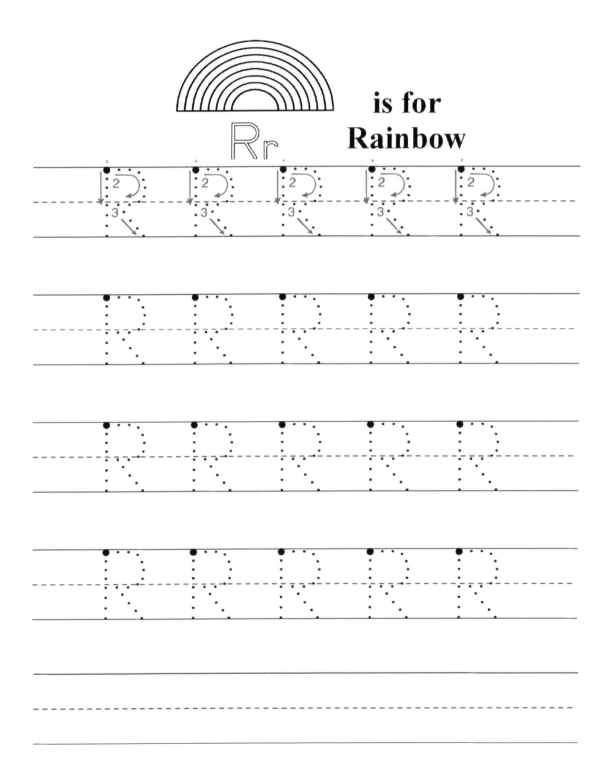

is for
Rainbow

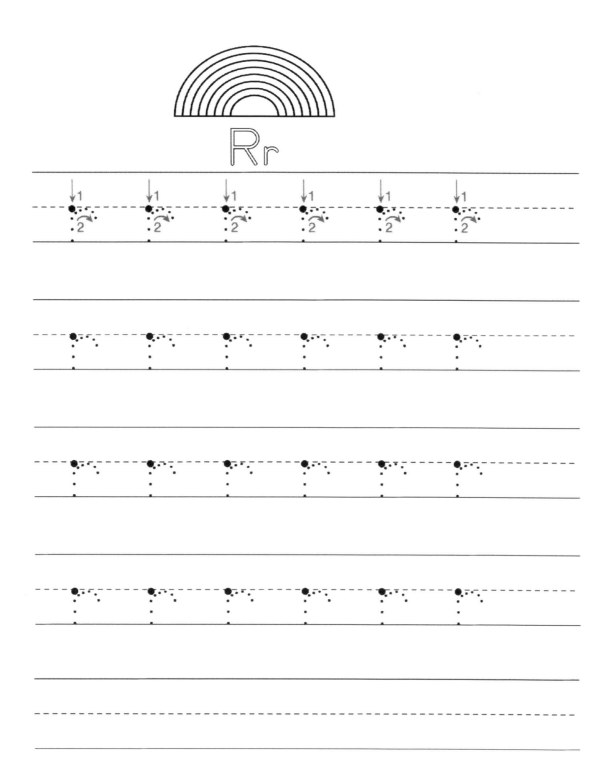

Rr

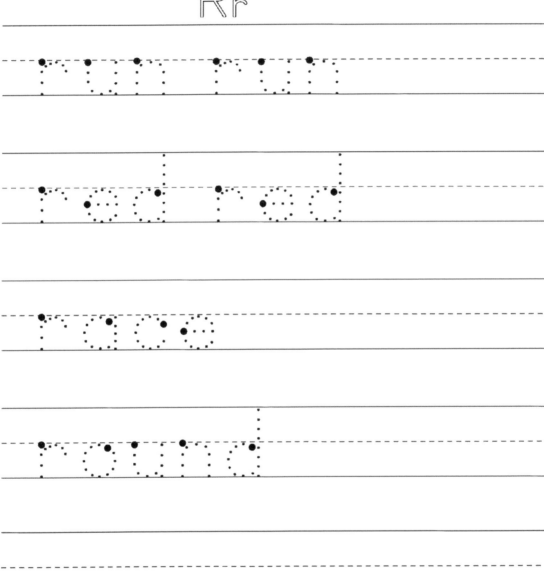

run run

red red

race

round

is for
Stool

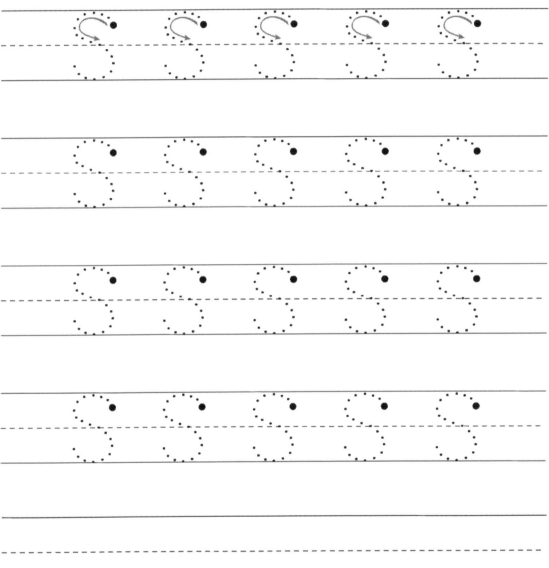

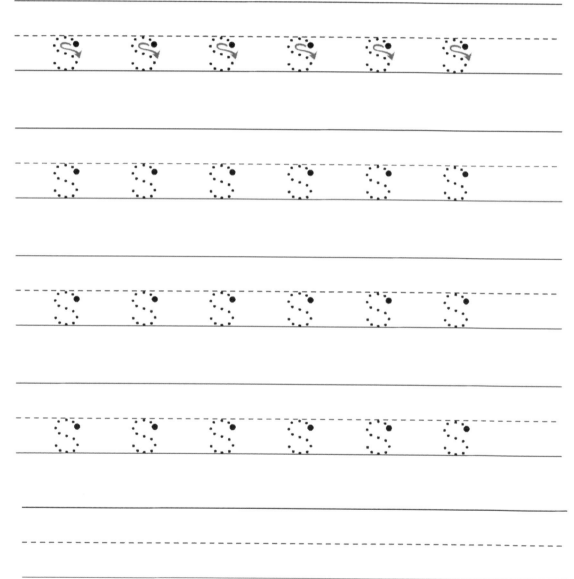

so so

sun sun

some

such

T t is for Teapot

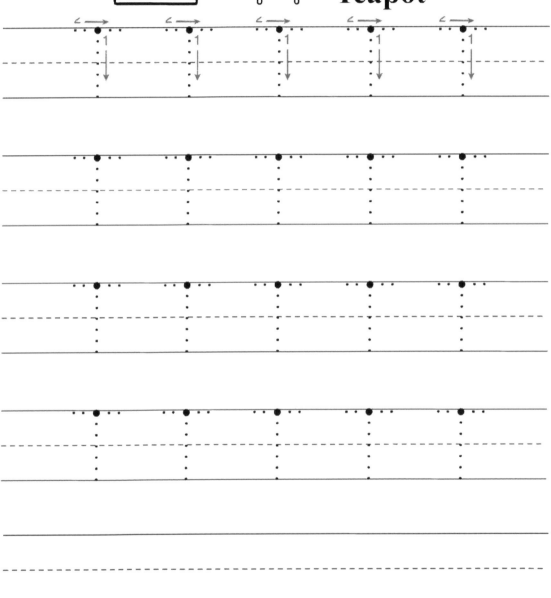

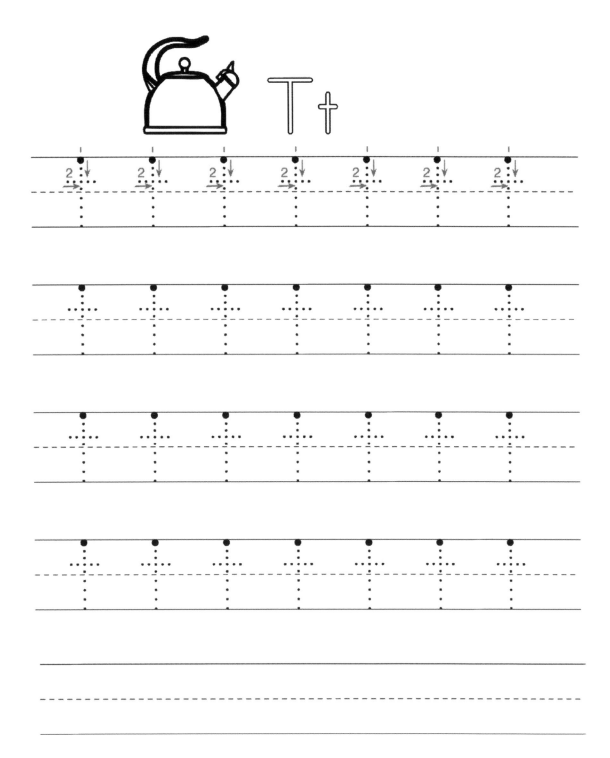

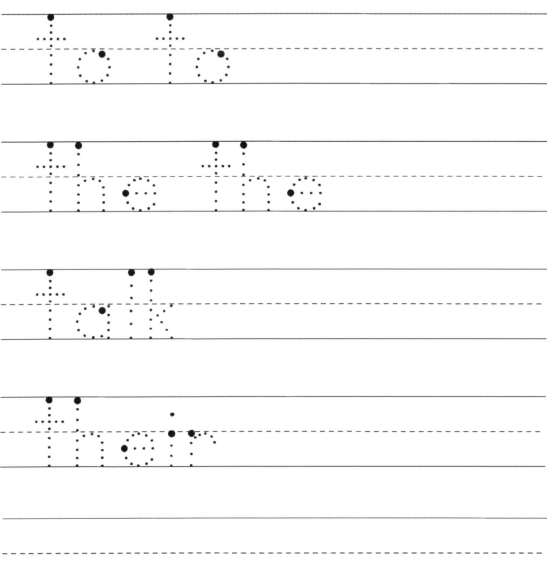

to to

the the

talk

their

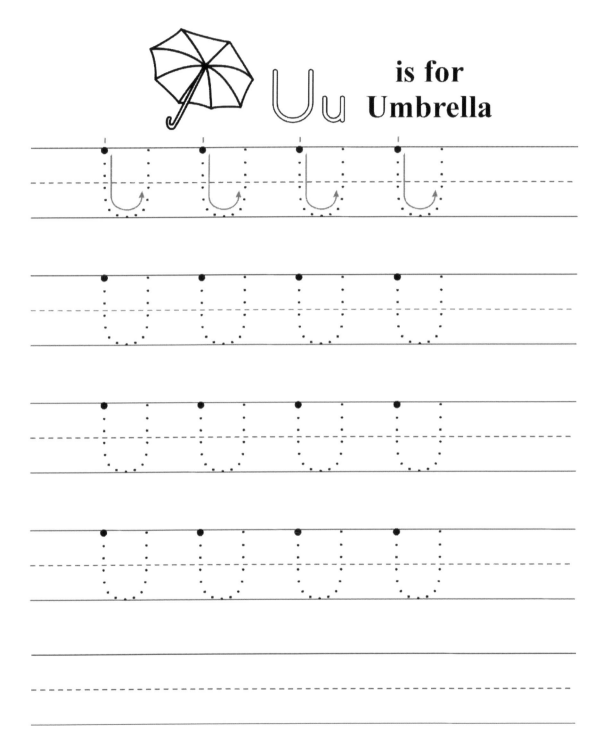

is for
Umbrella

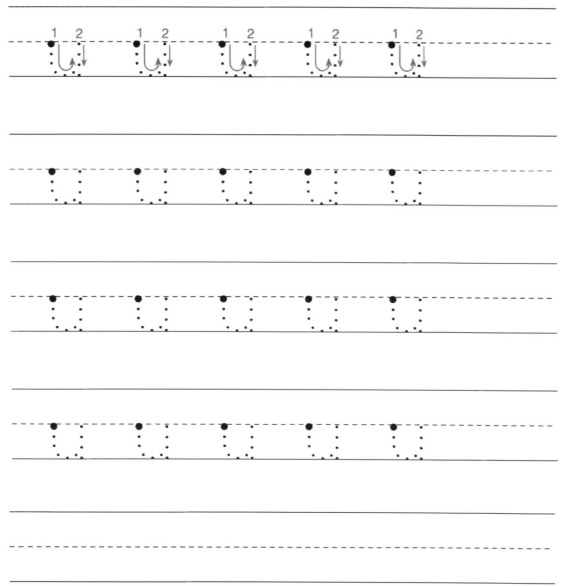

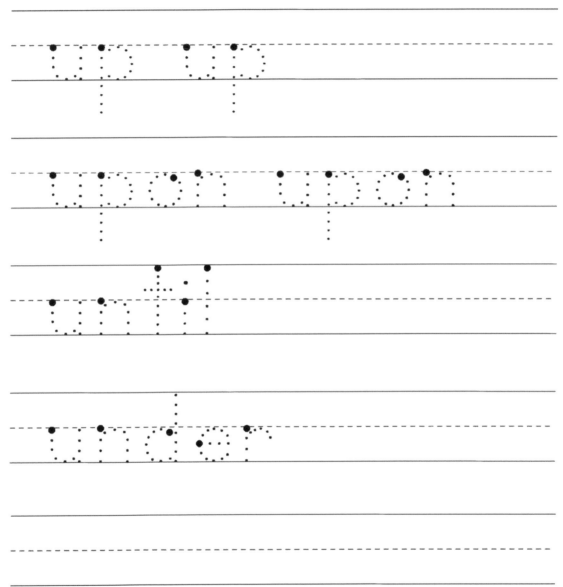

up up

upon upon

until

under

is for
Vest

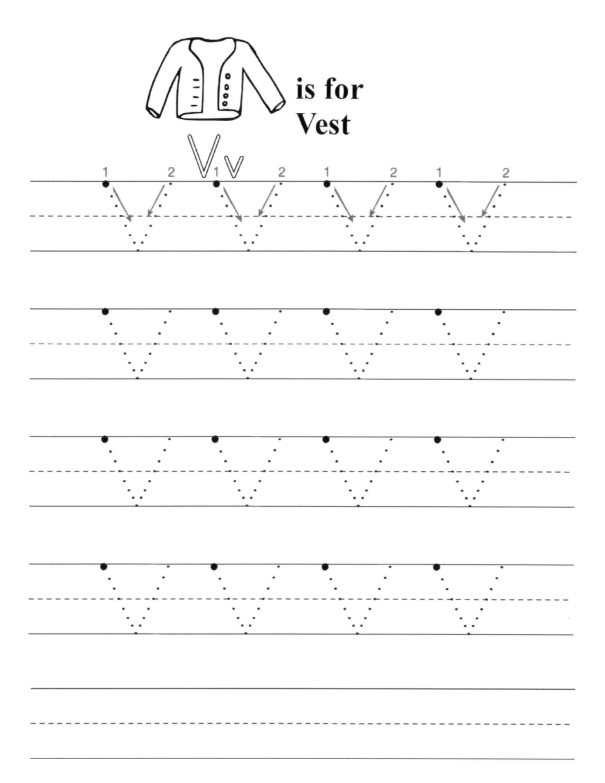

V v

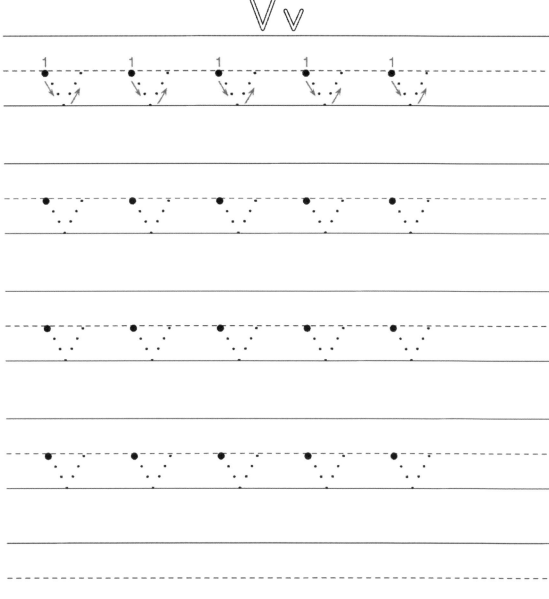

V v

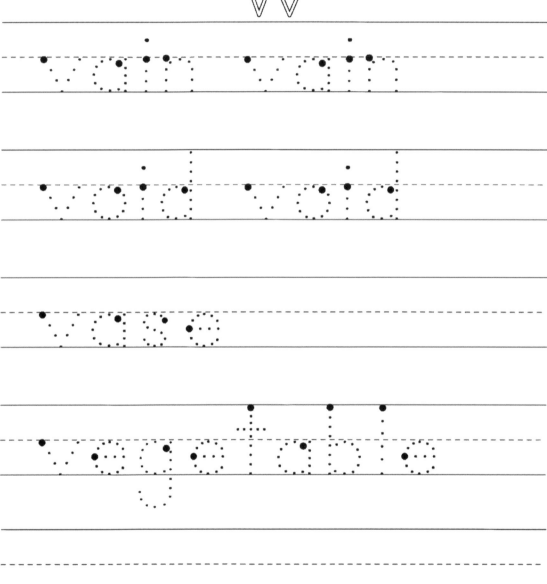

vain vain

void void

vase

vegetable

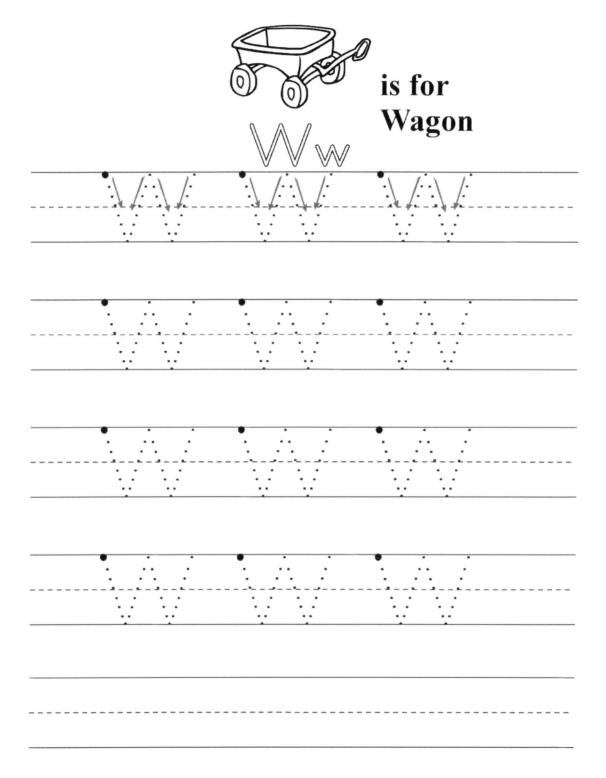

is for
Wagon

Ww

Ww

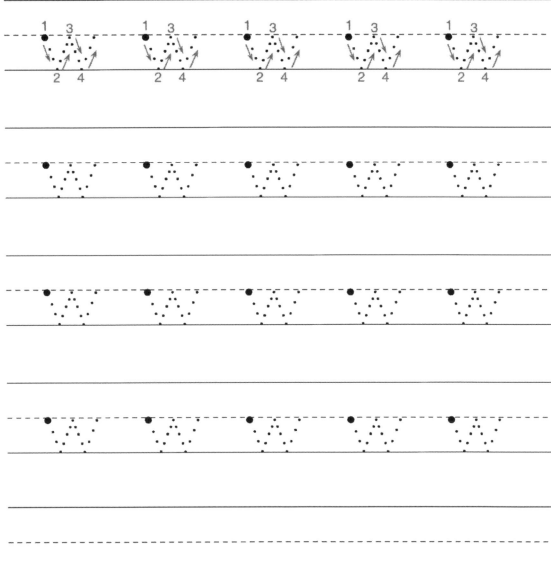

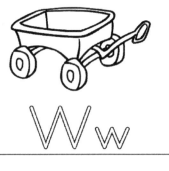

Ww

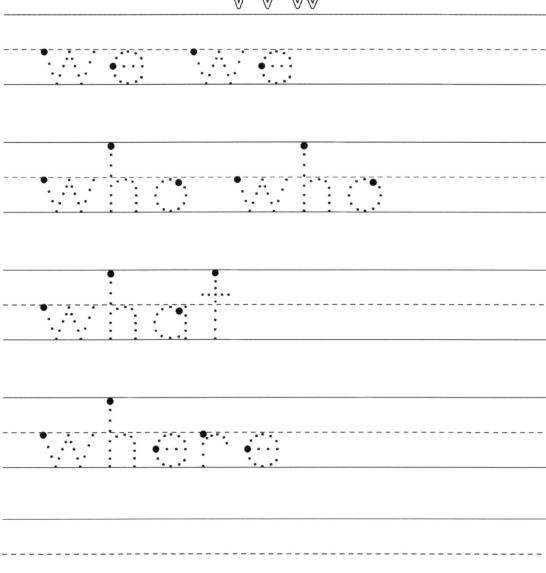

we we

who who

what

where

is for
Xylophone

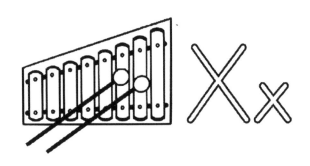

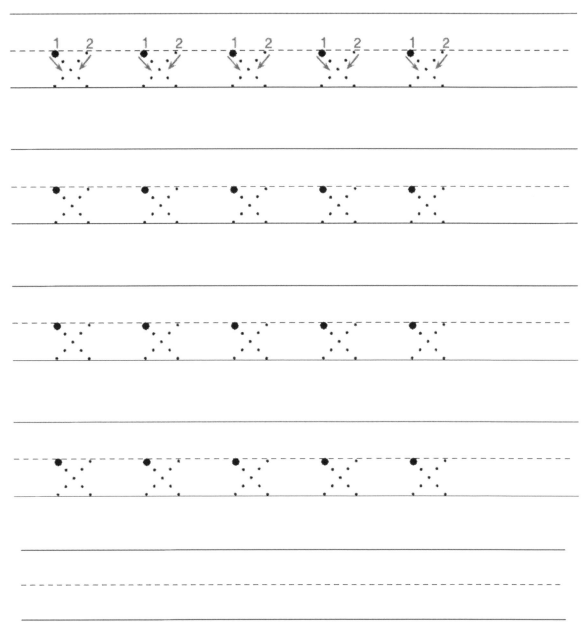

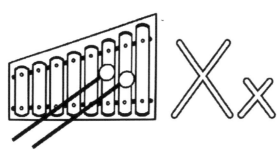

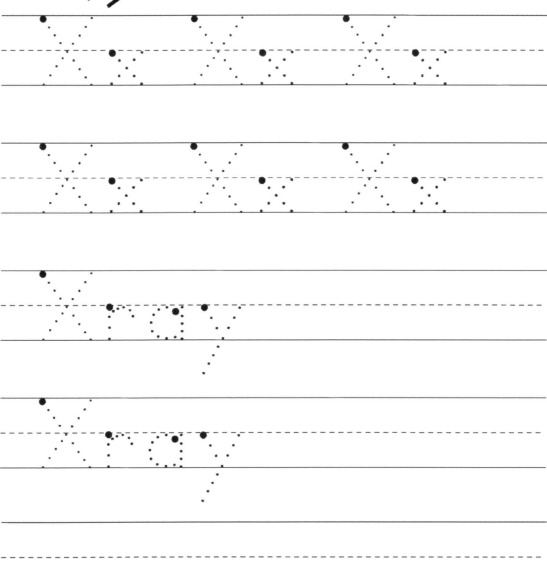

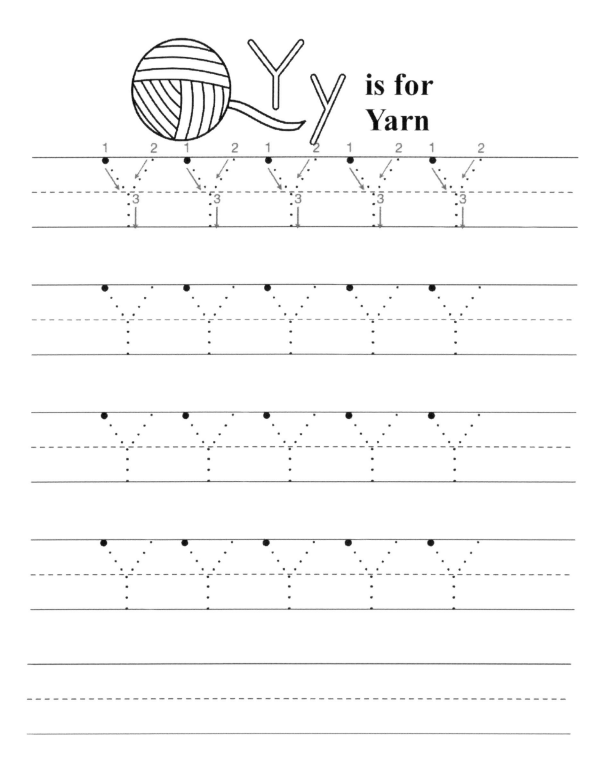

Yy is for Yarn

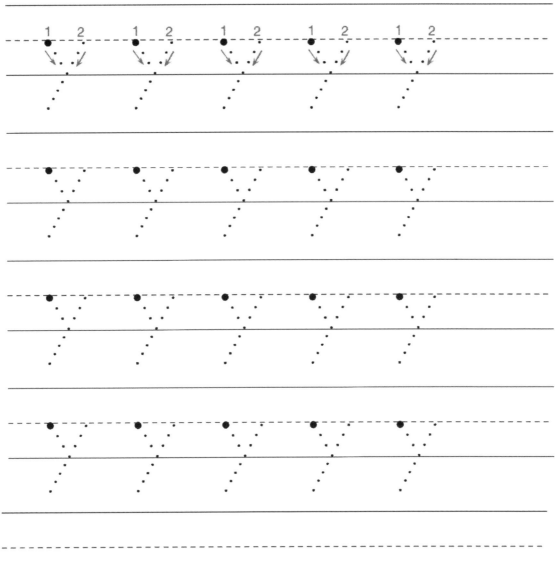

you you

yes yes

your

year

**is for
Zebra**

Z z

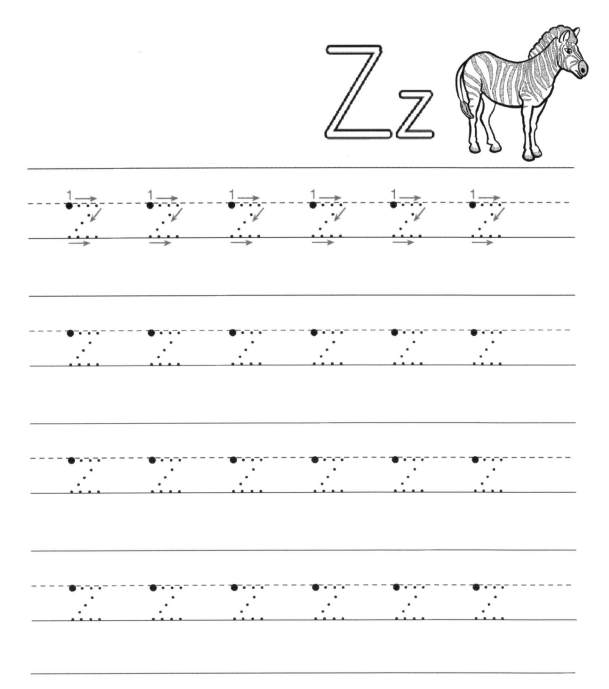

Zz

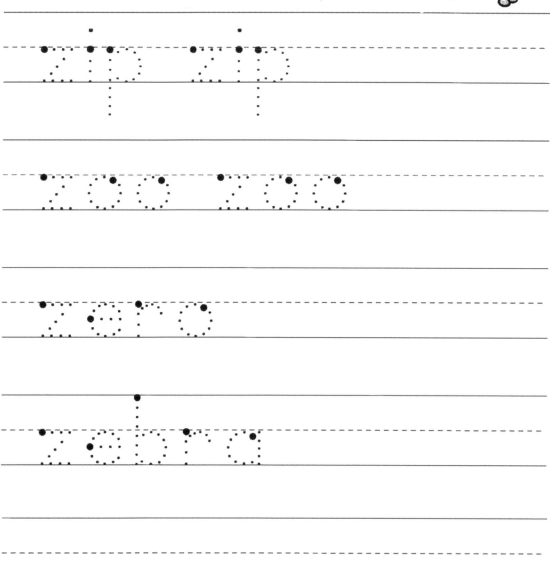

zip   zip

zoo   zoo

zero

zebra

69780296R00062

Made in the USA
San Bernardino, CA
21 February 2018